AF228363

INSIDE AREA 51

TOP SECRET

ISAAC KERRY

Lerner Publications ◆ Minneapolis

Lerner Publications Company
An imprint of Lerner Publishing Group, Inc.
241 First Avenue North
Minneapolis, MN 55401 USA

For reading levels and more information, look up this title at www.lernerbooks.com.

Main body text set in Aptifer Sans LT Pro.
Typeface provided by Linotype AG.

Editor: Brianna Kaiser **Designer:** Athena Currier

Library of Congress Cataloging-in-Publication Data

Names: Kerry, Isaac, author.
Title: Inside Area 51 / Isaac Kerry.
Description: Minneapolis : Lerner Publications, [2023] | Series: Top secret (alternator books) | Audience: Ages 8–12 | Audience: Grades 4–6 | Summary: "Area 51 is a secretive location that has led to many conspiracy theories. Examination of those theories, the location's history, and more will keep readers engaged as they discover the mysteries of Area 51"— Provided by publisher.
Identifiers: LCCN 2022006211 (print) | LCCN 2022006212 (ebook) | ISBN 9781728476612 (library binding) | ISBN 9781728478326 (paperback) | ISBN 9781728485348 (ebook)
Subjects: LCSH: Area 51 (Nev.)—Juvenile literature. | Air bases—Nevada—Juvenile literature. | Unidentified flying objects—Juvenile literature.
Classification: LCC UG634.5.A74 K47 2023 (print) | LCC UG634.5.A74 (ebook) | DDC 001.942—dc23/
    eng/20220222

LC record available at https://lccn.loc.gov/2022006211
LC ebook record available at https://lccn.loc.gov/2022006212

Manufactured in the United States of America
1-52242-50682-6/20/2022

TABLE OF CONTENTS

Introduction: Mysteries in the Desert 4

Chapter 1: **A Secret Base in America** 6

Chapter 2: **Planes and Flying Saucers** 12

Chapter 3: **The Future of Area 51** 22

Timeline 29

Glossary 30

Learn More 31

Index 32

INTRODUCTION

MYSTERIES IN THE DESERT

High above the Nevada desert, a glowing object soars through the sky. It flies higher than any aircraft could. Someone sees it land deep in the desert and attempts to follow it to its landing site. After a while, they stop. Before them stands a fence with barbed wire, armed guards, and signs reading RESTRICTED AREA—NO TRESPASSING. They're not sure of what they saw. But they're unlikely to find out because the object landed in Area 51, a secret military base in the US.

Area 51, a US Air Force base, is 38,400 acres (15,540 ha) of land in Groom Lake, Nevada. Groom Lake is a salt flat,

or dry lake. The salt flat and surrounding mountains are fenced in. Area 51 has been the home to many top-secret projects. Armed guards and helicopters patrol it. Only people with clearance are allowed in the base. This secrecy has led to many different conspiracy theories. Some can be explained. Others remain mysteries.

Area 51 has been connected to flying saucer sightings.

CHAPTER I

A SECRET BASE IN AMERICA

Warning signs on a fence outside Area 51

Some theories say crop circles near Area 51 were created by aliens.

The beginning of Area 51 lies in the Cold War (1947–1991), a conflict between the US and the Soviet Union. The Soviet Union was made up of Russia and fourteen other countries. They were Communist, a kind of political system. The US wanted to stop the spread of communism, creating a conflict between them and the Soviet Union.

The US and the Soviet Union spied on each other during the Cold War. They each had their own spy force. The US had the Central Intelligence Agency, or CIA, since 1947.

In 1954 the CIA wanted to create an advanced spy plane that the Soviets could not stop. Building and testing such a secret aircraft required a highly secure location. The CIA picked Groom Lake. The large salt flat worked as a runway.

This map shows which countries aligned with one another during the Cold War. The USSR and its allies supported communism, while the US and its allies did not.

A CIA building in Virginia in 1961

Groom Lake was also next to the government's atomic bomb testing range, so security was already in place. The base was soon staffed with guards, test pilots, and engineers. Over the years, the number of employees expanded. They worked on some of the most secret projects in the country.

A sign near the atomic bomb testing range

Leading to Mystery

The CIA has declassified only some details of Area 51's history. The declassified information helps explain why the base has always been cloaked in secrecy. While this secrecy managed to keep many of Area 51's projects hidden from the public, it also created mystery. Over the years, many people have shared conspiracy theories about Area 51. They try to explain what the government is doing in its remote desert base. Many of these theories feature a common element: aliens.

CHAPTER 2

PLANES AND FLYING *SAUCERS*

To some people, apparent flying saucer sightings are proof of alien life.

Local sheriffs guard one of Area 51's entrances in 2019.

Conspiracy theories about aliens and Area 51 have been around for decades. These theories are popular for many reasons. One reason is the research into advanced spy planes that happened at Area 51. Another is a mysterious crash that happened in Roswell, New Mexico, in 1947. Most important is the government's extreme secrecy about the base.

A U-2 spy plane can appear as a glowing object during flight.

Spying at 70,000 Feet

Beginning in 1954, the CIA worked on developing the world's most advanced spy plane. They eventually developed the U-2 aircraft. The U-2 was ahead of its time. It could fly at 70,000 feet (21,336 m). This was higher than any Soviet planes could fly. This meant that the Soviet Union could not stop the U-2 during its missions.

The U-2 was coated in titanium. This expensive metal is extremely strong and lightweight. It is also very reflective.

As the planes flew, the sun reflected off the bottom of their wings, creating the appearance of a glowing object shooting through the sky. This caused numerous reports of unidentified flying objects, or UFOs, around Area 51.

The CIA was worried about these reports. It did not want its spy plane program discovered. If the Soviet Union learned the details of their design, they would be able to create countermeasures. And the CIA had planned for the planes to fly

A U-2 spy plane in 1960

over Soviet airspace. Soviet knowledge of the spy planes could create tension between the countries and even start a war.

The US government was also very concerned about the public's fear of UFOs. In 1938 a radio station in New Jersey staged a radio drama of *The War of the Worlds*. This science fiction novel by H. G. Wells tells the story of an alien attack on Earth. The radio station did such a good job with their radio drama that people thought it was real and panicked. This alarmed the CIA. They were worried reports of UFOs could cause mass panic.

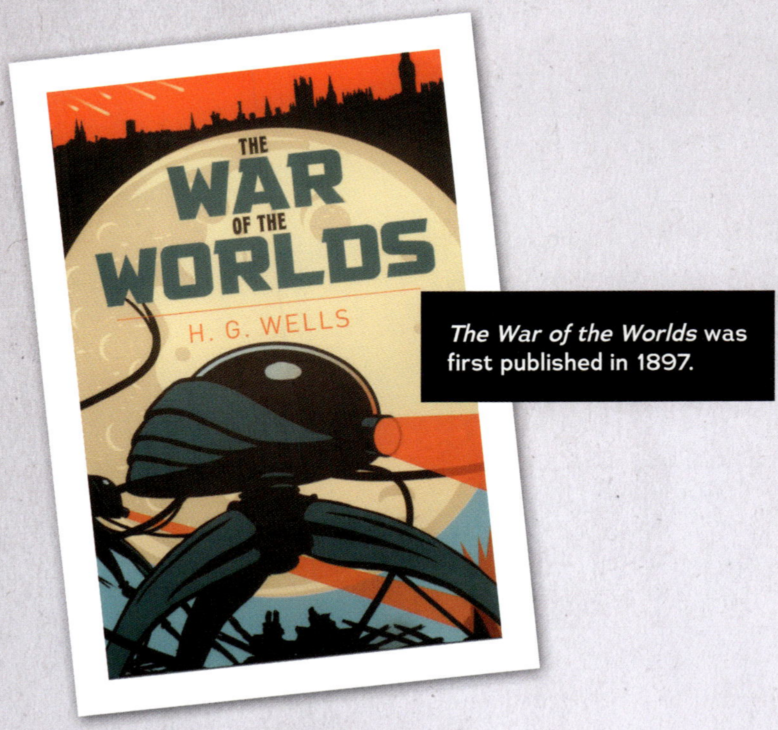

The War of the Worlds was first published in 1897.

In October 1938, Orson Welles (*center*) talks to reporters about the radio drama he directed and narrated.

Secrets in the Desert

In July 1947, something crashed in a field outside of Roswell, New Mexico. The nearby Roswell Army Air Field was notified. An army press officer, Walter Haut, released a statement to the local radio station. He said the US Army Air Forces found a flying disc near Roswell.

Only three hours after Haut's statement, he returned to the radio station. This time, his message was very different. He said his first statement had been incorrect. He claimed the object that had crashed was just a weather balloon.

Many people believe that Roswell, as well as Area 51, is connected to aliens.

The Roswell crash did not cause much public interest at the time. It would take over thirty years for it to gain national attention. In 1978 Stanton Friedman, a physicist and UFO researcher, and his research partner, Bill Moore, learned of the Roswell crash and decided to investigate. They found numerous witnesses who claimed to remember the events from thirty years ago. Witnesses said they saw a flying saucer. Then they said the military swarmed the town after the crash and put large boxes and crates onto military trucks. Many witnesses even claimed to have seen small bodies that didn't look human removed from the wreckage.

A July 1947 newspaper headline about the Roswell crash

Stanton Friedman (*right*) speaks at an event in 2013.

Testimonies given thirty years after an event are problematic. Memories can fade and change. The portrayal of aliens in the years after the crash could also affect what people remember. Even though the information they got from witnesses could have been false memories, Moore wrote of his and Friedman's research in the 1980 book *The Roswell Incident*. Roswell became a key example of the theory that the government covers up UFO activity.

In 1989 a former employee of Area 51 claimed that the government was studying alien spacecraft there. Conspiracy theorists quickly made the connection to the wreckage recovered in Roswell. Area 51 was now associated with aliens and UFOs. The previous rumors of strange flying objects seemed to confirm this.

Historical events can have many different stories about what happened. With the Roswell crash, there are conflicting stories of a weather balloon crash and a flying saucer. How would you try to discover the truth? What strategies could you use? You might interview eyewitnesses, compare published information, or maybe even go to New Mexico!

Debunking UFO Sightings

In 1952 the CIA responded to the rise in UFO sightings and encouraged the public not to worry. They started a campaign to debunk UFO sightings. In 1993 the CIA declassified some documents about the campaign, but most of the details are still classified. People's personal experiences with UFOs and government secrecy both contribute to creating alien conspiracy theories, as well as other theories, about the base.

CHAPTER 3

THE FUTURE OF AREA 51

An SR-71 Blackbird at the Edwards
Air Force Base in California

The F-117 Nighthawk bomber is just one
type of plane developed at Area 51.

Area 51 continued to be an important research
facility throughout the Cold War. In addition to creating the
U-2, engineers at Area 51 developed stealth technology and
other advanced aircraft, including the SR-71 Blackbird spy
plane and the F-117 Nighthawk bomber. These planes could
avoid being detected by enemy radar.

The US National Guard uses a Black Hawk (*right*) and Chinook helicopter (*left*) to help put out a forest fire in 2013.

DECLASSIFIED

One conspiracy theory is that the US government has a secret force of black helicopters. Depending on who tells the story, the helicopters have different purposes. Some believe they are a special force to deal with UFO events, while others believe they are a secret military force preparing to take over the US. But there is another explanation.

Area 51 has a small fleet of helicopters that operates out of the base. They provide security around the base and do local search and rescues. These helicopters are sometimes called Ghost Hawks. In 2011 a Navy Seal raid in the Middle East revealed the existence of highly secretive military technology: stealth helicopters. Sources claimed that these aircraft were developed and tested at Area 51.

The MQ-9 Reaper is a type of unpiloted aircraft.

Area 51 has also worked on developing drones. These unpiloted aircraft were intended for missions during the Cold War. As missile technology increased, more and more of the CIA's spy planes were shot down. By using unpiloted drones instead, the CIA could still fly dangerous missions but not risk any pilots.

Advances in weapon technology eventually allowed drones to be equipped with missiles. They are now a critical part of the air force's fleet. Rumors are that the researchers at Area 51 are working on the next generation of unpiloted combat aircraft.

Area 51 continues to be associated with aliens and government conspiracies. In 2019 a Facebook group called Storm Area 51, They Can't Stop All of Us formed. The group wanted to find out what the government may be hiding at Area 51, including information about aliens. While not many

A 2016 satellite image of Area 51

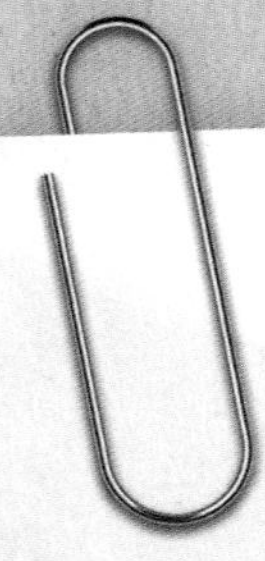

SOLVE IT

Highly advanced technology always has been a key part of Area 51's story. Can new technology help solve some of the base's mysteries? Maybe! One possibility may be commercial drones. They can hold cameras and take photos inside the base. Over time, it will become harder to keep them, and other technology, out of places such as Area 51.

people showed up for the raid, it shows that the public still thinks of Area 51 as a mysterious site involved with otherworldly things.

Many of these conspiracy theories have their roots in historical facts. However, until the government decides to declassify all its records about Area 51, we will never know what has taken place at the secretive US base.

Timeline

1947: A mysterious object crashes in a field in Roswell, New Mexico, and leads to conspiracy theories.

1954: A remote site is selected in the Nevada desert for the CIA's secret spy plane program.

1956: The CIA's advanced spy plane, the U-2, can fly at 70,000 feet (21,336 m).

1980: *The Roswell Incident* is published. Public interest in the Roswell crash grows.

1989: A former employee of Area 51 claims to have seen alien ships at the base.

1993: The CIA declassifies some documents about a campaign to debunk UFO sightings.

2011: A Navy Seal raid reveals that the US has stealth helicopters.

2019: A Facebook group called Storm Area 51, They Can't Stop All of Us forms.

Glossary

aircraft: a vehicle designed to fly through the air

airspace: the air above a nation that counts as part of it

associate: to be connected to something

clearance: official permission to access a location or information

conspiracy theory: a theory that explains a fact or event as being caused by a secret plot

countermeasure: a technology designed to make an opponent's weapon ineffective

debunk: to prove something is false

declassified: to officially make a piece of information no longer secret

stealth: a way of designing an aircraft to allow it to be undetected by radar

trespass: to enter a place unlawfully

Learn More

Area 51: Britannica Kids
https://kids.britannica.com/students/article/Area-51/631585

Boutland, Craig. *The Roswell UFO Mystery*. Minneapolis: Lerner Publications, 2020.

Karst, Ken. *Area 51*. Mankato, MN: Creative Education, 2021.

Manzanero, Paula K. *Where Is Area 51?* New York: Penguin Workshop, 2018.

O'Keefe, Emily. *Investigating UFOs*. New York: AV2 by Weigl, 2020.

Roswell Incident: Britannica Kids
https://kids.britannica.com/students/article/Roswell
-Incident/313285

The Roswell Report: Air Force
https://www.af.mil/The-Roswell-Report/

What Exactly Are UFOs?: CBC Kids
https://www.cbc.ca/kids/articles/what-exactly-are-ufos

Index

aircraft, 4, 8, 14–15, 23, 25–27
aliens, 11, 13, 16, 20–21, 27

Central Intelligence Agency (CIA), 7–8, 11, 15–16, 21, 26
classified information, 21
Cold War, 7, 23, 26
conspiracy theories, 5, 11, 13, 20–21, 25, 27–28

declassified information, 11, 21, 25

Friedman, Stan, 19–20

Groom Lake, NV, 4, 8–9

helicopters, 5, 25

military, 4, 19, 25
Moore, Bill, 19–20

Roswell crash, 13, 17, 19–21
Roswell Incident, The (Berlitz and Moore), 20

unidentified flying objects (UFOs), 15–16, 19–21, 25
U-2, 14–15, 23

War of the Worlds, The (Wells), 16

Photo Acknowledgments

Image credits: DigitalGlobe/Getty Images, pp. 1, 27; Fer Gregory/Shutterstock, p. 5; BRIDGET BENNETT/AFP/Getty Images, p. 6; Feng Cheng/Shutterstock, p. 7; Bettmann/Getty Images, p. 8; AP Photo/ Henry Burroughs, p. 9; Ted Streshinsky/ CORBIS/Getty Images, p. 10; Obsidian Fantasy Studio/Shutterstock, p. 11; PhonlamaiPhoto/Getty Images, p. 12; George Frey/Getty Images, p. 13; John Bryson/Getty Images, p. 14; Heritage Space/Heritage Images/Getty Images, p. 15; Ben Molyneux/Alamy Stock Photo, p. 16; AP Photo, p. 17; chrispecoraro/ Getty Images, p. 18; M L Pearson/Alamy Stock Photo, p. 19; Abaca Press/Alamy Stock Photo, p. 20; Stocktrek Images/Getty Images, p. 22; Avpics/Alamy Stock Photo, p. 23; US Army Photo/Alamy Stock Photo, p. 24; US Air Force Photo/Alamy Stock Photo, p. 26.

Design elements: Marjan Blan/unsplash.com; fotograzia/Getty Images; Reddavebatcave/Shutterstock; AVS-Images/Shutterstock.

Cover image: Fer Gregory/Shutterstock.com; fotograzia/Getty Images.